Orlando   Austin   New York   San Diego   Toronto   London

Visit *The Learning Site!*
www.harcourtschool.com

Introduction

On a hot September day, I parked my rented car in a lot high above the Arizona side of Hoover Dam. The dam is enormous. Its top is wide enough to support two lanes of traffic inching back and forth between Arizona and Nevada. Fed by the Colorado River, the blue waters of Lake Mead seemed trapped behind the dam, which is 221 meters (726 ft) high. But I knew differently. Some of the lake's waters run through the dam.

After walking across the dam to the Nevada side, I entered the visitors' center and joined a group of tourists. A tour guide led us into a large elevator. Motors hummed, and we began to drop.

Soon the elevator came to a smooth stop, and we filed out. I heard a mysterious droning sound. After a few seconds of walking, I found myself looking over a vast room. The droning had grown louder. Now I knew where the sound had come from.

Water flowing through generators within Hoover Dam produces electricity.

Huge electric generators groaned beneath me. Somewhere under those generators the flowing waters of Lake Mead turned the blades of turbines that drove the generators. The generators, and the elevator motors that brought us to them, were made possible by the discoveries of two long-dead scientists. Here are their stories.

Hans Christian Oersted

Few people have heard of the little town of Rudkoebing on the Danish island of Langeland. Yet most people who have studied electricity and magnetism have heard of the man who was born there on August 14, 1777.

Soeren Oersted was a pharmacist in the town of Rudkoebing when his older son Hans Christian was born. A year later, in 1778, another boy, Anders, was born. Both brothers were bright and inquisitive. However, neither of them received a formal education while growing up in Rudkoebing. Because of family problems, when they were still very young they were sent to live with a German wigmaker and his wife. It was there that they learned German, Latin, French, and basic math. Other people in the town also recognized the brothers' great thirst for education and did what they could to teach them.

When he was about 11, Hans began working in his father's pharmacy. There he took an early interest in chemistry and the sciences.

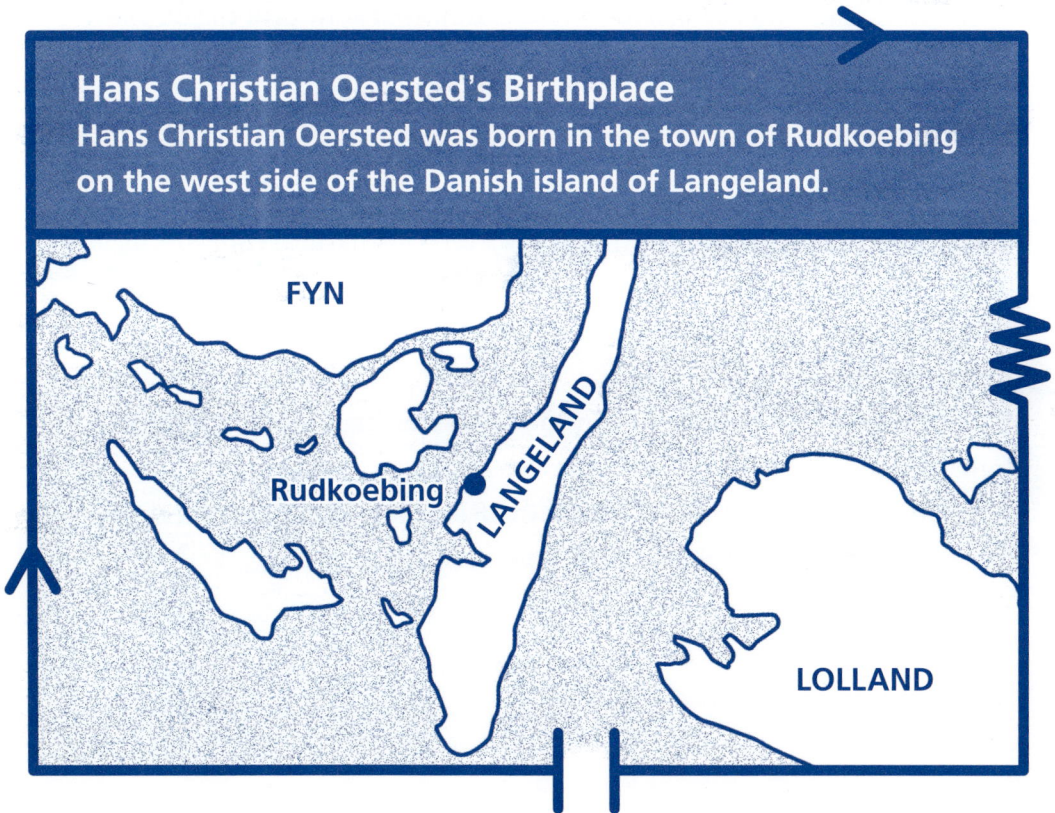

Hans Christian Oersted's Birthplace
Hans Christian Oersted was born in the town of Rudkoebing on the west side of the Danish island of Langeland.

A World of Wonders

In spite of their unstructured education, both brothers were accepted into the University of Copenhagen in 1794. Hans studied astronomy, chemistry, mathematics, physics, and, of course, pharmacy science. Anders studied law and eventually became one of Denmark's finest lawyers. He also became the attorney general of Denmark and, eventually, prime minister.

In the meantime, Hans's scientific studies had made him very curious about sciences other than pharmacy science. He was especially curious and excited by a discovery made by the Italian scientist Alessandro Volta. Volta had discovered how to produce a steady electric current.

Some scientists believed that electricity and magnetism were related. Others laughed at the idea, but no one seemed to be able to find an answer to this question. It was a question that Hans, too, pondered during his studies and much later into his career.

In 1801 Hans was awarded a scholarship that sent him to Germany and France. During his three years there, he met other scientists who believed in a philosophy of nature called Romanticism. One aspect of this philosophy was that all forces are unified. During this period of his life, his reputation became somewhat damaged when he recklessly accepted faked experimental evidence that supported this natural philosophy.

In 1806 Hans returned to the University of Copenhagen, where he became a physics professor. He taught there for six years, and then he traveled for two years. During his travels, as he met again with scientists in Germany and France, he began to reconsider his opinions about Romanticism.

Hans returned to the university in 1814. Then one day something unexpected happened that changed the history of science.

A Needle Twitches

On an evening in April 1820, Oersted was lecturing a group of students about electricity. His lecture dealt with electric currents. As part of his lecture, he passed an electric current through a wire. There was nothing special in that; scientists had done this many times.

Hans Christian Oersted at work in his laboratory.

However, on this particular day, something other than an electrified wire lay nearby on the laboratory table. It was a compass.

As Oersted demonstrated his electrified wire, he passed it over the compass. It is not known whether Oersted did this deliberately or by accident. But there is no doubt about what then happened. The magnetized compass needle twitched!

None of the students seemed to have seen the twitch. Oersted was sure he had seen it. Had the electric current in the wire caused the twitch? If so, Oersted had found a connection between magnetism and electricity. But no scientist relies on an accident to prove a point. Oersted knew that he had to perform a carefully designed experiment to duplicate what he thought he had seen.

When Oersted was ready to perform his experiment, he connected a wire to a battery. Electricity flowed through the wire. He moved the wire over the needle of a large compass. The needle didn't budge. Had his eyes betrayed him back in that April classroom?

Oersted went over what he had just done. He had moved the wire across the needle at a right angle to it. What would happen if he moved the wire in a direction parallel to the needle?

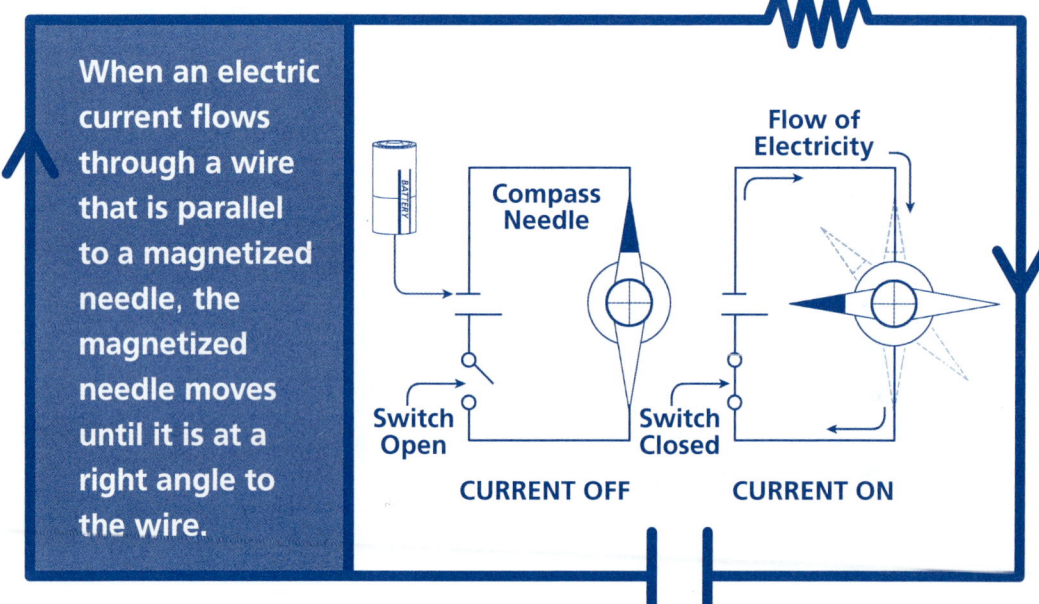

When an electric current flows through a wire that is parallel to a magnetized needle, the magnetized needle moves until it is at a right angle to the wire.

He tried this. The needle jerked and turned to face the wire! Oersted had caused motion with electricity. In other words, he had shown that an electric current can make a magnet move. The only way this could happen was if the electric current produced a magnetic force to which the needle responded.

This discovery led to the invention of the electric motor. Electric motors contain magnets that spin an electromatic bar when the coil wrapped around the bar is exposed to electricity. Such motors are used in kitchen appliances, electric carpenter's tools, and elevators, like the one that took me down to the depths of the Hoover Dam.

An electric motor is made of a stationary permanent magnet and an iron bar that can move. When electricity in your home flows through coils around the iron bar, the bar becomes magnetized. It becomes an electromagnet. The bar is made to move because the like poles of the permanent magnet and the electromagnet repel each other. House current, or AC current, keeps changing direction, so the poles of the electromagnet keep changing. This makes the electromagnet spin continuously.

Electric Current (AC)   Electric Current (AC)   Electric Current (AC)

Electromagnet   Permanent Magnet

Michael Faraday

The Faraday family arrived in London, England in 1804. They had traveled from Newington, a small town about 30 miles to the east. James Faraday had been a blacksmith in the town making iron shoes for horses.

But an economic depression had hit the area. It was hard for James to make a living, and his health was not good. James hoped that the bustling city of London might give him more opportunities to earn a decent living.

In those days, the children of a poor family had to go to work at an early age. That left little time for school, so there was not much hope that the Faraday children would get a decent education.

Soon after arriving in London, James's son Michael Faraday found work running errands for a maker and seller of books by the name of Riebau. Michael was 14 years old when he stumbled into the world of books. These unfamiliar objects would become a solid road for young Michael to journey on. A year later, Riebau promoted Michael to apprentice bookbinder. (An apprentice is someone who begins to learn a trade on the job.) But Michael would leave his mark in history in an altogether different trade.

Treasures on the Pages

During his years at Riebau's shop, Michael did more than bind books. "There were plenty of books there, and I read them," he later told a friend.

One evening, when work was slow, Michael browsed through a copy of the *Encyclopedia Britannica.* Among the pages of the great book, he found a 127-page description of electricity. He took notes and made drawings of experiments in electricity and the setups used to make them. This sparked Michael's desire to learn more about electricity and science.

The bound notes lay among other books in Riebau's store. Then one day one of Riebau's customers, William Dance, entered the store. He looked through one book after another. Finally he came upon Faraday's bound notes. Only a very good student of science could have made the notes, he thought. He asked Riebau who had made the notes. Riebau pointed to Faraday.

Dance asked Faraday whether he would like to attend a series of lectures at the Royal Institution. Faraday was excited to hear that the lecturer would be Sir Humphry Davy.

Davy was then one of England's greatest scientists. Among his most famous accomplishments was using electricity to break chemicals into their separate parts. In this way, Davy discovered the elements potassium, sodium, barium, and a number of other metals.

On February 29, 1812, Faraday attended the first of Davy's four lectures. As Davy spoke, Faraday scribbled pages and pages of notes. He missed nothing.

Faraday was now 21 years old and needed to choose a career. Should he become a bookbinder, something he had learned to do very well? Or should he try the risky path toward science, something he knew little about? Science easily won out. But Faraday wondered how to start.

Faraday's first step was to bind the notes he had taken during Davy's lectures. The bound notes ended up in several volumes. Faraday sent the volumes to Davy along with a note asking for a job at Davy's laboratory. Unfortunately, there was no job available at that time. However, not long afterward, one of Davy's assistants was fired, and Davy offered the position to Faraday. Faraday began his life as a scientist. It is said that he worked very hard for a much smaller salary than he earned as a bookbinder. But it was his chosen career, and that was what mattered most to him.

Michael Faraday as a young man.

Getting to Work

For seven months, Faraday worked in Davy's laboratory—cleaning equipment. Then in October 1813, Faraday set out with Davy and his wife on a tour of Europe. For a year and a half, Faraday traveled in Europe, meeting some of the finest scientists of the time. He finally returned home in April 1815.

Now he was ready to perform experiments of his own. In 1820 Faraday read an article written by a little-known Danish scientist—Hans Christian Oersted. The paper told how Oersted had used electricity to produce magnetism.

This triggered a question in Faraday's mind. Could the reverse be done? In other words, could magnetism be used to produce electricity? In 1822 Faraday wrote in a notebook that he wanted to make an attempt to convert magnetism into electricity.

But for nine years other responsibilities prevented Faraday from trying this experiment. For one thing, he became head of the Royal Society's laboratories. The Royal Society was then England's top scientific institution.

Faraday worked hard at his new job, but he never forgot what he had wanted to do in 1822. Faraday was 40 years old when he finally was able to attempt the experiment. The date was August 29, 1831.

On that day, Faraday wound a wire around half of an iron ring (A). He wound a second wire around the other half of the ring (B). He connected the ends of the wire that formed coil B together. Faraday placed a length of this wire over a needle that had been magnetized, like the needle of a compass.

Faraday then connected the ends of the wire that formed coil A to a battery. Electricity then flowed through this wire. The needle under the coil B wire twitched. The twitch could have been caused only by electricity flowing through the coil B wire. Yet this wire was not attached to a battery or to the coil A wire; it was attached only to itself. How could electricity be flowing through the coil B wire?

Faraday searched for an answer to this question. Then it came to him. When he passed electricity through the coil A wire, the A half of the iron ring became magnetized. This produced a magnetic field. The magnetic field moved outward and crossed the coil B wire. As it moved over the coil B wire, the magnetic field produced an electric current in the coil B wire. In this way, magnetism produced electricity. Oersted had produced magnetism from electricity, and now Faraday had produced electricity from magnetism.

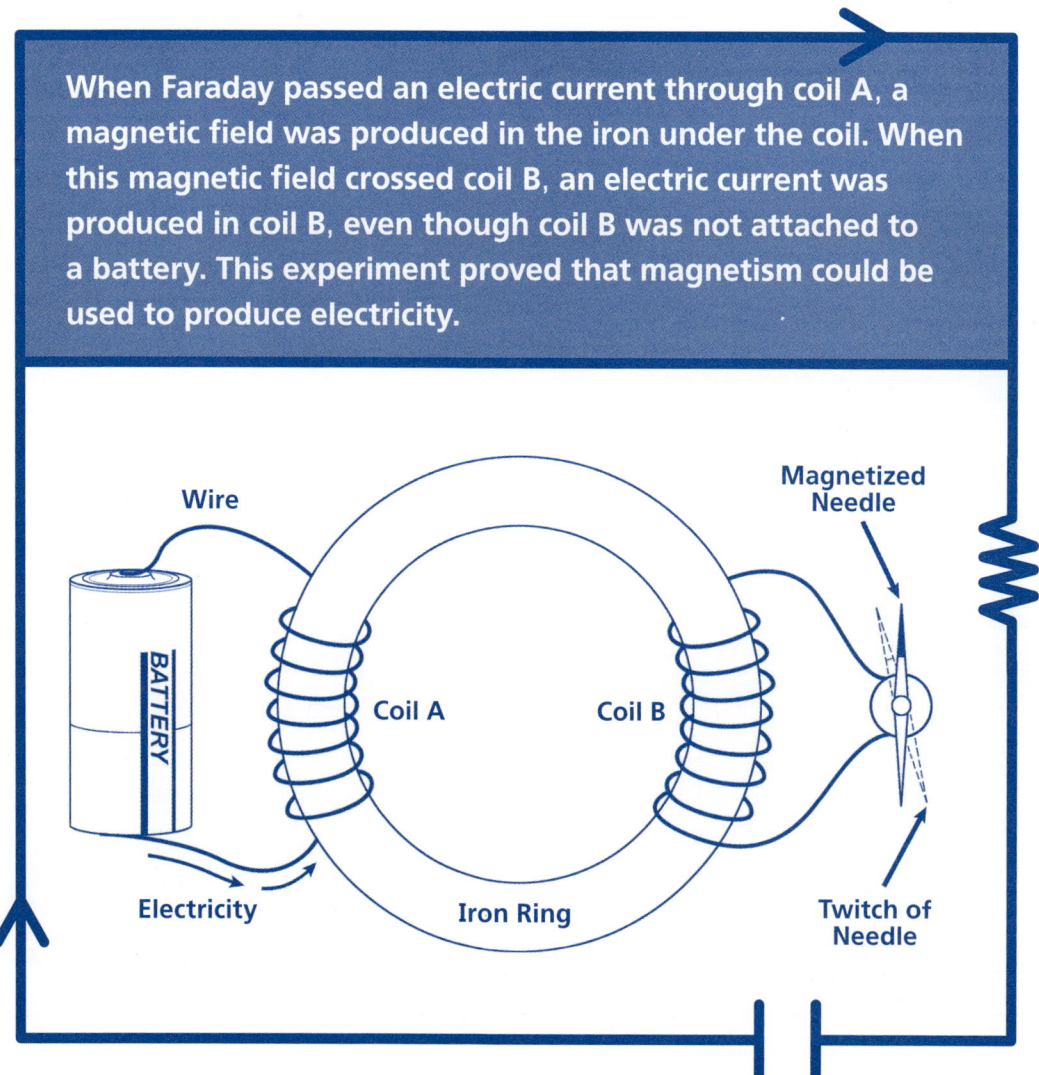

When Faraday passed an electric current through coil A, a magnetic field was produced in the iron under the coil. When this magnetic field crossed coil B, an electric current was produced in coil B, even though coil B was not attached to a battery. This experiment proved that magnetism could be used to produce electricity.

Oersted's discovery had led to the invention of a useful device, the electric motor. Faraday's discovery led to another useful device, the electric generator.

Electric generators, like those in the Hoover Dam, use magnetism to produce electricity. The process starts with the energy of motion and ends with the energy of electricity. In the Hoover Dam, the motion comes from the waters of the Colorado River that move through the dam. These moving waters turn the blades of a machine called a turbine. The blades are attached to a shaft that leads to a generator. Coils of wire cover the shaft of the generator. The shaft and wires turn between the poles of a huge magnet. As in Faraday's experiment, the magnet produces an electric current in the wire. The electricity travels out of the generator and eventually through power lines to homes, factories, schools, and cities.

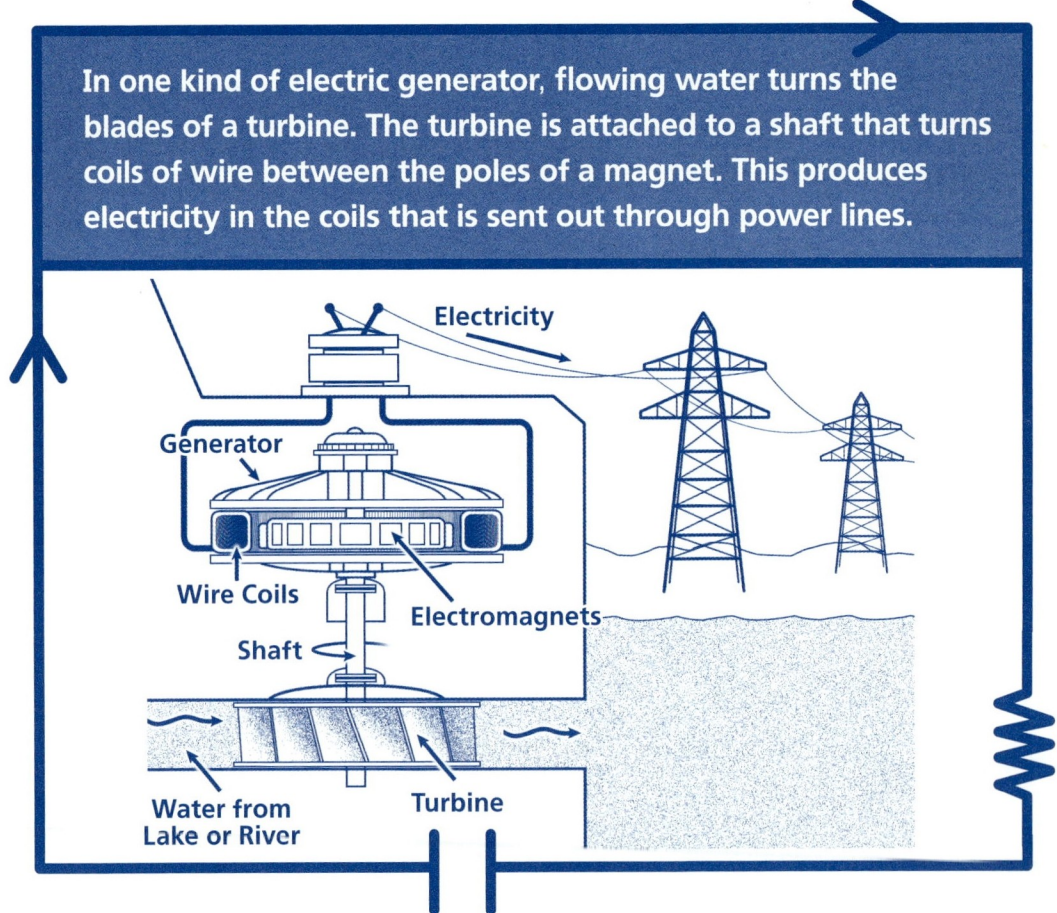

In one kind of electric generator, flowing water turns the blades of a turbine. The turbine is attached to a shaft that turns coils of wire between the poles of a magnet. This produces electricity in the coils that is sent out through power lines.